Addison Wesley
Science & Technology 5

Matter and Materials
•
Changes in Matter

Steve Campbell   Jim Wiese

Douglas Hayhoe   Beverley Williams

Doug Herridge   Ricki Wortzman

Lionel Sandner

Addison Wesley

Toronto

Coordinating & Developmental Editors
Jenny Armstrong
Lee Geller
Lynne Gulliver
John Yip-Chuck

Editors
Susan Berg
Jackie Dulson
Christy Hayhoe
Sarah Mawson
Mary Reeve
Keltie Thomas

Researchers
Paulee Kestin
Louise MacKenzie
Karen Taylor
Wendy Yano, Colborne Communications Centre

Consultants
Dr. Ron Kydd, Professor and Head of Chemistry, University of Calgary
Lynn Lemieux, Sir Alexander MacKenzie Sr. P.S., Toronto District School Board

Pearson Education Canada would like to thank the teachers and consultants
who reviewed and field-tested this material.

Design
Pronk&Associates

ISBN 0–201–64986–1

This book contains recycled product and is acid free.
Printed and bound in Canada.

3 4 5 – TCP – 04 03 02 01

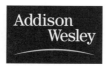

Changes in Matter

What do you have in common with a tree? What about a cloud or a peanut butter sandwich? All the things you see, touch, and taste are made of matter. This includes trees, clouds, peanut butter sandwiches, and you! In fact, anything that takes up space and has mass is matter.

But some things, like trees, are hard. Some things, like water, are wet and flow. And some things, like clouds, float in air. All these things are made of matter, but they exist in different forms, or states. Sometimes a type of matter will change its state from one form to another. This unit will show you why changes in matter are important to us.

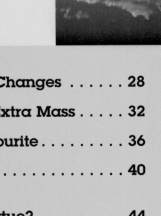

Now you will find out:

- what are the three states of matter
- what is a change of state
- how changes of state influence the designs of devices
- about the properties of everyday materials and how using them affects the environment

What's the Matter?

Get Started

You're about to walk into a new type of museum. It's called the Interactive Technology Museum of Arts and Sciences. This museum lets you be part of the action! Take a look at this picture to see the different rooms throughout the building. The Grand Opening theme is "What's the Matter?"

STATES OF MATTER ROOM

ALWAYS CHANGING ROOM

Grand Hall

Work On It

Work with a partner. Describe each room in the building. Can you guess what is the main idea of each room? Next, make a list of all the changes you see happening.

Communicate
Discuss Write

1. Share your ideas about the rooms with your classmates. Record all your ideas on the blackboard.

2. Take the list of changes you made. Put the changes into groups that make sense to you. Give a title to each group.

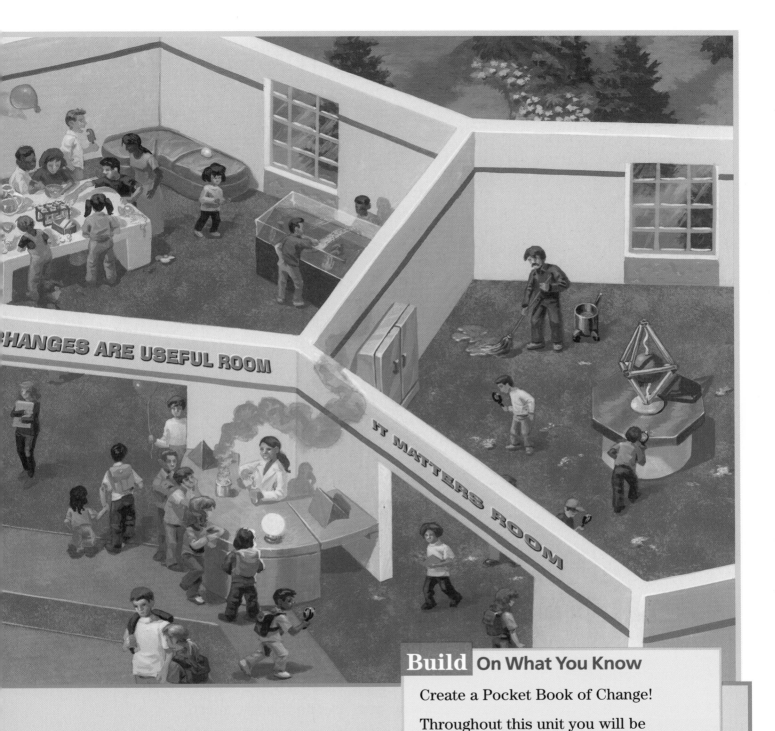

CHANGES ARE USEFUL ROOM

IT MATTERS ROOM

3. When you get home, look around. Do you see any changes occurring? Make a list of any changes you see.

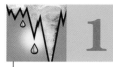

I am in every picture. Sometimes I am the same, sometimes I am different. What am I?

Get Started

After passing through the Grand Hall, you peer inside the first room. It seems a bit confusing. You notice three pictures on the wall with a riddle written below them. Try to answer the riddle. Share your thoughts with your classmates.

Work On It

You probably have an answer to the riddle of the three pictures. Now it's time to look more closely at your answer. In the following activities, you will collect additional information that will help you explain what's happening in the three pictures.

The Strange Science Mix

You will work with your teacher to complete this first activity.

Materials for each class:

60 mL of water   spoon

charcoal briquette   60 mL of salt

small bowl or beaker

60 mL of liquid bluing

15 mL household ammonia

small clear plastic container

Procedure

1. Watch carefully as your teacher mixes the liquid ingredients together to make a "science mix." What happens?

2. Next, your teacher will put the charcoal briquette in a clear plastic container and pour the mixture over it. The science mix should just cover the briquette. What happens?

3. Record how much mix your teacher put in the container.

4. Draw a labelled diagram of what you see. Include the date, time, and the height of the mixture in the container.

5. Draw a second diagram that predicts what you think the container will look like in one week.

6. Your teacher will put the container in a safe spot where it will not be disturbed. Check the container each day and record any changes you observe.

What Happens to Ice?

Materials for each group:

safety goggles   2 or 3 ice cubes

wooden spoons or other tools for breaking up the ice

Procedure

1. Put on safety goggles.

2. Predict what you expect will happen to ice when you hold it in your hand.

3. Use the wooden spoons to break the ice cubes into pieces.

4. Take a piece of ice in your hand and hold it. What happens?

Water Changes

Part One

Materials for each group:

foil plate   3 ice cubes

250 mL beaker

boiling water (your teacher will give this to you later)

Procedure

1. Place the ice cubes on the foil plate.

2. Put the 250 mL beaker on a flat surface where it will not tip over.

3. Ask your teacher to pour boiling water into your beaker. The beaker should be about half full of water.

4 Place the foil plate with the ice cubes on top of the beaker.

5 Watch the ice cubes on the plate for a few minutes. What happens?

6 After a few minutes, carefully lift the foil plate 10 cm above the beaker. Look at the bottom of the plate. What do you see?

Part Two

Materials for each group:

250 mL beaker   water

masking tape

Procedure

1 Put some water into the beaker so it is about $\frac{1}{4}$ full.

2 Put a piece of tape down the side of the beaker.

3 Draw an arrow on the piece of tape to show the level of the water.

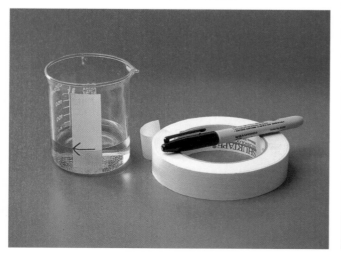

4 Put the beaker in a safe, warm place. Leave it there until the next day.

5 Check your beaker the next day. What happened to the water level?

Communicate

Write   Discuss

1. What do you think is the answer to the three-picture riddle? How did your investigations help you to understand your answer better? Discuss your answer with your group.

2. What caused the ice to melt in the What Happens to Ice activity?

3. Where do you think the water on the bottom of the foil plate came from, in Part One of the Water Changes activity? How did it get to the plate?

4. What do you expect to happen to the level of water in the beaker in Part Two of the Water Changes activity? Explain your answer.

5. What **properties**, or characteristics, does ice have that are not shared by liquid water?

A property of something is what it looks like or how it behaves.

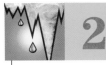

2 What's the State?

Get Started

In the last activity you probably solved the riddle by noticing that each picture had water in it. Water comes in many different forms: **solid**—like snow and ice, **liquid**—like water in a lake, and **gas**—like steam from a boiling kettle, or **water vapour** in clouds or fog. These different forms are called **states**. Can you think of any other substance that you have seen in three different states?

As a matter of fact, water is the only material that we can normally see on Earth in the form of a solid, liquid, and gas.

In the next activity you will rotate to different stations. At each station, you will explore a variety of different materials to determine the characteristics of each state.

Work On It

Matter can exist in three states: solid, liquid, and gas. How can you tell the difference between these states? Try to answer the following questions about the state of some materials as you perform the activities below.

- Can the shape of this material change?
- Can the **volume** of this material change? Remember, the volume of something is how much space it takes up.

Station 1—Rock Solid

Materials at this station:
A selection of different-shaped solids like a rock, pencil, penny, golf ball, wooden block, etc.

Procedure

1 Look at the objects in front of you. Can their shape be changed easily?

2 Can you change the volume of any of these objects?

3 Name some of the properties that all of these objects have in common.

Station 2—Sloshing Liquid

Materials at this station:
graduated cylinder

beaker (about $\frac{1}{4}$ full of water with food colouring in it)

Procedure

1 Pour the coloured water into the graduated cylinder. Measure and record the volume of water.

2 Pour the water into the beaker again and then back into the graduated cylinder.

3 What is the volume of the water now?

4 Did the shape of the water change at any time during steps 1 and 2?

5 Did the volume of the water change at any time during steps 1 and 2?

Station 3—Pressing Water and Gas

Materials at this station:
Syringe (50 mL to 100 mL)

100 mL beaker ($\frac{1}{2}$ full of water)

Procedure

Part A: Water

1 Put the tip of the syringe under the surface of the water in the beaker. Slowly pull the plunger out until the syringe is $\frac{1}{2}$ full of water. Try not to let air bubbles in.

2 Cover the tip of the syringe with your finger so no water leaks out. Try to push the plunger into the syringe. How far can you push the plunger?

Safety Caution

Be careful not to break the syringe.

3 Did the shape of the water change as you pulled it into the syringe?

4 Did the volume of the water change in step 2?

Part B: Air

1 Empty all the water from the syringe.

2 Half-fill the syringe with air by pulling the plunger halfway out.

3 Cover the tip of the syringe with your finger as you did before. Try to push the plunger into the syringe. How far can you push the plunger?

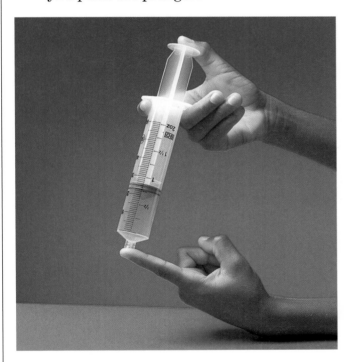

4 Did the shape of the air change as you pulled it into the syringe?

5 Did the volume of the air change in step 3?

Station 4—What's the State of Sugar?

Materials at this station:

sugar cubes   magnifying glass

paper towels

wooden block or something else to crush the sugar cubes

Procedure

1 Describe the shape of the sugar cube. How could you calculate its volume?

2 Place the sugar cube in a paper towel and crush the cube.

3 Pour the grains onto a dark surface.

4 Observe the grains carefully using your magnifying glass. Draw the shape of the sugar grains you observed.

5 What would happen if you crushed the sugar even more?

6 Is the crushed sugar a solid or a liquid? Explain.

Communicate
Discuss   Write

1. Copy the table below into your notebook. Fill in the table by describing some of the characteristics of the three states, based on your observations from this activity.

State	Shape	Volume
Solid		
Liquid		
Gas		

2. Copy the table below into your notebook. Fill in the table by identifying each material as a solid, liquid, or gas. Include an explanation for each.

Material	State	Reason for answer
Margarine in a fridge		
Butter in a warm frying pan		
Smoke		
Dish soap		
A quarter		
Lemonade		
Steam from a boiling kettle		
A milkshake		
A shoe		
Cooking oil		
A Popsicle		

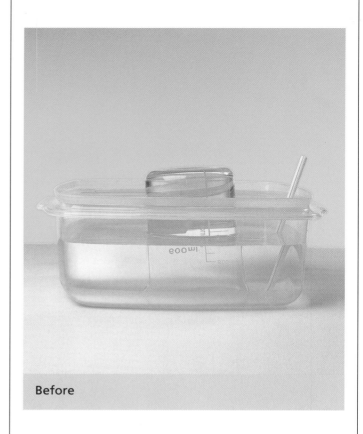

Before

3. A beaker full of water is placed upside down in a tub of water. After placing a flexible straw under the beaker, a student blows through the straw.

a) Discuss with your class what is happening in these pictures.

b) What happened to the water in the beaker? Where did it go?

c) What replaced the water in the beaker? Where did it come from?

d) What characteristic of air did you observe in these pictures? Discuss this question with your class.

4. Why is it good that solids do not change their shape easily? Write a paragraph explaining what your life would be like if all solids changed their shape and volume easily.

After

3 Changes of State

Get Started

In the last activity you had an opportunity to explore the different properties, or characteristics, of the three states of matter. Everything that has mass is made of matter. You are made of matter, and so are all the things you see around you—and even some things you don't see, like gas.

List the three states of matter on a blank sheet of paper. This is your States of Matter Sheet. For each state, give an example of something you observed in Lesson 2. If you have trouble, pair up with a classmate and share your ideas.

Next, look at these pictures. Identify the state of each thing shown. When you are done, share your ideas with your classmates. If someone comes up with a good example for a particular state, add it to your sheet using a different-coloured pen or pencil.

Now you have identified the three states of matter. All matter you experience in life exists in one of these three states. Do you remember which is the only material that we can see on Earth in all three states? Use the photos below to remind yourself.

60% of our bodies are made of water.

We use water to cook.

Two-thirds of Earth's surface is covered by water.

Weather involves water in all three states—rain, snow, and clouds.

An average of 375 icebergs flow south of Newfoundland each year.

The air you breathe contains water in the form of water vapour.

If you said water, you're right. Water is a special substance that is critical to us and Earth's survival—we can't live without it. As a solid we call it ice, as a liquid we call it water, and as a gas we call it steam, or water vapour.

Changes of State

When a substance changes from one state to another, it has gone through a **change of state**. Changes of state occur when heat energy is added or taken away from a material. We often see changes of state of water. For example, an ice cube sitting on your desk will melt as the warm air in your classroom adds heat to the ice cube.

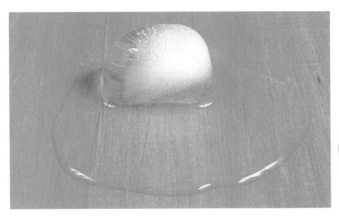

What if you take a can of pop out of the fridge and leave it on the table? The sides of the can will get wet! This is another change of state, as the water vapour in the air changes to liquid water on the sides of the cold can.

Here are four common changes of state.

Change of State	What happens
Melting	Solid changes to a liquid
Evaporation	Liquid changes to a gas
Condensation	Gas changes to a liquid
Freezing	Liquid changes to a solid

Now, go back to the States of Matter Sheet you started at the beginning of this lesson. Add these four changes of state to your sheet using a different-coloured pen or pencil than you used before. Include an example for each. Look at your observations from the Water Changes activity in Lesson 1 for more examples of changes of state.

 If Time Allows

Can you think of any more examples of changes of state? Include them on your States of Matter Sheet.

Communicate
Write

1. Name the change of state that is occurring in the following situations.

   a) The ice on a pond begins to crack and disappear as the temperature warms up.

   b) When a kettle boils, steam comes out of the spout.

   c) After a candle has been lit, wax begins to drip down the sides.

   d) You make Popsicles by putting a mould full of juice into the freezer.

2. Fill in the blanks of the following short story.

A _____ gently floated down from the sky on a cold winter evening. It lay on the ground in the _____ state until spring. Then, as the temperature warmed up, the snowflake _____, changing into the _____ state. It turned into _____ and flowed into a puddle at the bottom of the yard. Soon, the temperature warmed up even more and the town became very hot. The drop of water in the puddle _____ and changed into the _____ state. It was now _____. Moving into the sky, the water vapour floated around and eventually cooled, _____, and changed into the _____ state. It fell back to Earth as a _____. The water cycle had been completed once again and was starting another loop.

3. Review your States of Matter Sheet. Name one thing you didn't know before that you know now.

4. How are changes of state useful? Give an example of a useful change of state in your life.

Build **On What You Know**

Add to your Pocket Book of Change a page that shows the three states of matter and the four changes of state discussed in this lesson. Include a picture or example of each state and each change of state.

4 Keep It Hot— Keep It Cold

Get Started

Imagine you are going to the beach on a hot summer day. There's no food stand, so you need to pack a lunch—but how will you keep your food and drinks cool on such a hot day?

Now imagine a cold winter day. You're going out to play in the snow with your friends. It sure would be great to take some hot chocolate with you—but how can you keep your hot chocolate warm? What can you do to solve these problems?

Work On It

Like most people, you probably use devices that keep your hot food hot and your cold food cold. Hot food, like hot chocolate, gets cold when it loses heat. Cold food, like ice cream, gets warm when it gains heat. Devices that stop food and other things from losing heat or gaining heat are called **insulating devices.**

Look at these pictures of insulating devices. What do you think they have in common that helps them stop the loss or gain of heat? Discuss your thoughts with a partner and record your ideas. Don't lose these ideas—you're about to put them to good use!

In these activities, you will use the materials you are given to design insulating devices that can keep things hot or cold.

Keep It Hot

Your task is to keep a cup of warm water as close to the starting temperature as possible for 10 min. Hint: your device should stop heat from escaping.

Materials for each pair:

plastic drinking cup	warm water
newspaper	thermometer
masking tape	Styrofoam

a lid for the cup (the lid should have a hole in it for the thermometer)

any other materials your group may choose

Procedure

1 Choose the materials your group thinks will be best to build your insulating device.

2 Draw a picture or diagram of the device you are going to make.

3 Build your device. Leave the lid of the cup off—your teacher will put it on later.

4 Once you have built your device, your teacher will add warm water to your cup, and attach the lid.

5 Put the thermometer through the hole in the lid into the water. Record the starting temperature of the water.

6 Use a table like the one started below to record the temperature every minute for 10 min.

Time (min)	Temperature (°C)
0 (starting time)	
1	
2	
3	

7 Copy the following graph into your notebook. Graph your results by plotting the temperatures you recorded at each minute.

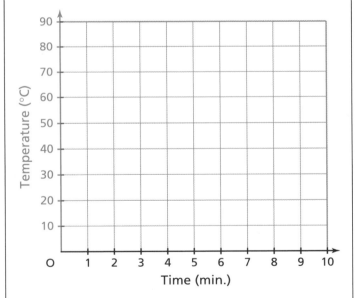

Keep It Cold

The following note has come from your principal.

To: The Grade 5 Science Class
From: The Principal

I have just had a call from the owner of a local pet store. The store is about to get a delivery of a tiny Antarctic animal. The store has very little information about this animal, except that it needs to live in a very cold, icy environment. It will take a while before the store owner can create a proper habitat for this animal. She needs to find a way to design a container that is $\frac{1}{2}$ full of ice for the animal. The ice must keep for as long as possible without melting. She would like some suggestions. I immediately thought of this class to help solve her problem.

Your task is to design and build a device to meet the request from your Principal.

- You can use any materials you choose, including the ones from the previous activity.
- Your device should include a plastic cup to keep the animal in.
- When your device is finished, half fill the cup with ice and put it in a safe place.
- Check your device every hour throughout the day until all the ice is melted.
- Can you keep the ice frozen for the entire day?

1. Look at your graph from the Keep It Hot activity. How did the temperature change over the 10 min? Try to use words like gradual change, or fast change, and temperature in your description.

2. a) Look at the insulating devices your classmates made in the Keep It Hot activity. Did anyone else's device keep the water warmer than your device?

   b) How could you improve your insulating device?

3. a) What was the temperature of the water in your cup after 1 min? What was the temperature after 5 min?

   b) Look at your graph from the Keep It Hot activity. What was the temperature in your cup after 8.5 min?

4. In the Keep It Cold activity, how long did it take for the ice in your device to melt?

5. Name some properties, or characteristics, of insulating devices that help them to slow down heat gain or heat loss.

6. a) What happens to a liquid, like water, if it loses a lot of heat (for example, if you put it in the freezer)?

   b) What happens to a solid, like ice, if it gains a lot of heat?

   c) What do insulating devices have to do with changes of state?

Build On What You Know

Check the results of the experiment your teacher demonstrated in Lesson 1. Record any changes you see in your Pocket Book of Change. Draw a picture in your Pocket Book to show what is happening. What do you think caused the changes you see? Check the experiment again in a few days.

5 Changes: Reversible or Non-Reversible?

MAGIC RAISINS

1. Put 100 mL of vinegar into a 600 mL beaker.
2. Add water to the vinegar up to the 400 mL level of the beaker.
3. Mix in two spoonfuls of baking soda.
4. Add a small handful of raisins.
5. Watch!

Get Started

The next room in the Interactive Technology Museum of Arts and Sciences is the Always Changing Room. You walk in and sit down at a table. On the table in front of you are some materials, and a computer with directions for a science experiment displayed on the monitor.

You try the activity. But very strange things start to happen! The mixture of vinegar, water, and baking soda is bubbling, and the raisins are dancing up and down in the beaker! What's going on? Discuss with your class what you think is happening here.

You have just read about some changes. In this activity, you will be observing lots more changes!

To help you keep track of your observations, start a Changes Observation Sheet. Take a blank sheet of paper and fold it in half three times. When you unfold the paper, you should have eight boxes. Draw lines in the folds so you can see the boxes more clearly. Use each box to record a change you see in these activities. To start, fill in a box with one or two changes that you think happened in the Magic Raisins activity described on the previous page.

Now you will investigate six different examples of change at six stations in your classroom. You should visit all six stations, in any order you wish.

Change Station 1

Materials at this station:

vinegar

paper towels

pennies

small containers

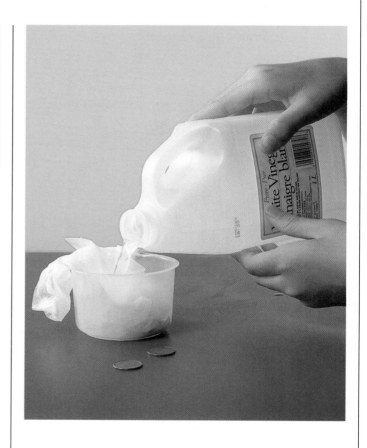

Procedure

1 Wrap two or three pennies in a paper towel. Put the paper towel with the pennies in the bottom of the container.

2 Pour some vinegar onto the paper towel. The paper towel should be very wet and there should be a small pool of vinegar in the bottom of the container under the paper towel.

3 Set the paper towel aside until the next day.

4 What do you think will happen? Why do you think so?

5 On your Changes Observation Sheet, use half a box to describe what you think will happen.

6 Check the pennies the next day. Use the other half of the box to describe the change you observe.

Change Station 2

Materials at this station:
antacid tablets
beaker of water

Procedure

1 Drop an antacid tablet into a beaker of water.

2 What changes do you see?

3 How long do these changes occur for?

4 Describe the changes you observed on one or two boxes of your Changes Observation Sheet.

Change Station 3

Materials at this station:
sugar
spoon
beaker of water

Procedure

1 Put a teaspoon of sugar into the beaker of water.

2 Stir it until all the sugar dissolves.

3 Where did the sugar go? Has it changed into something different, or is it still sugar? How do you know?

4 Describe the change you observed on a box of your Changes Observation Sheet.

Change Station 4

Materials at this station:
lemon juice
saucer
cotton swabs
blank paper
lamp

Procedure

1 Pour some lemon juice into the saucer.

2 Dip one end of a cotton swab into the lemon juice. Use it like a pen to "write a message" on the paper. Write slowly so there is enough lemon juice on each letter you write. Let the lemon juice dry.

3 Hold the paper close to a light. Make sure the paper doesn't touch the light. What do you see? What caused this change?

4 Describe the change you observed on a box of your Changes Observation Sheet.

Change Station 5

Materials at this station:

safety goggles   mini marshmallows

Plasticine   aluminum tart cups

paper clips

Procedure

1 Put on safety goggles.

2 Unwind the paper clip to make it straight.

3 Put a lump of Plasticine at the bottom of the aluminum tart cup.

4 Put a mini marshmallow on one end of the paper clip.

5 Push the other end of the paper clip into the Plasticine, so that it stands upright. Make sure the paper clip will not fall over.

6 Ask your teacher to light the mini marshmallow on fire. What do you see happening?

7 Describe the change you observed on a box of your Changes Observation Sheet.

Change Station 6

Materials for this station:

new steel wool

a set-up of steel wool in a beaker upside down in a see-through container of water (prepared by your teacher the week before)

a set-up of an empty beaker upside down in a see-through container of water

Procedure

Several days ago, your teacher placed some steel wool at the bottom of a beaker and put it upside down in a container of water. Your teacher also placed an empty beaker upside down in a similar container of water.

1 Look at the steel wool in the beaker. Compare it to a new piece of steel wool. What change do you observe?

2 Does this change make a difference in how useful the steel wool is? Explain.

3 Look at the water level in the upside down beaker with the steel wool. Compare it to the water level in the empty upside down beaker. What change do you observe?

4 Describe the changes you observed on one or two boxes of your Changes Observation Sheet.

 If Time Allows

Try the Magic Raisins activity from the beginning of this lesson. You could also try this at home. Use a see-through drinking glass instead of a beaker. Put vinegar in the glass so it is $\frac{1}{4}$ full. Then add water until the glass is almost full (about 2 cm from the top). Continue with the rest of the activity.

Communicate   Write Present Discuss

1. Use a pair of scissors to cut your Changes Observation Sheet into the eight equal boxes.

   a) Each box has a change described on it. Organize the eight changes into groups that make sense to you.

   b) Give each group a title.

   c) Pair up with a classmate. Look at each other's groups. Explain why you grouped the changes the way you did. You can change your groupings if you wish.

   d) Present your groupings to the class. Do your classmates agree with your groupings?

2. Which of the eight changes do you think are **reversible**, or can change back? Explain how you would change them back.

3. Which of the eight changes do you think are **non-reversible**, or cannot change back? Why not?

4. Identify and describe a change you observed that produced a gas.

5. List the four changes of state you learned about in Lesson 3. Can these changes be reversed or not? How do you know?

Build On What You Know

List the changes you observed in this activity in your Pocket Book of Change. Beside each change, explain whether you think it can be changed back or not. Don't forget to add what happens to the pennies in the paper towel with vinegar when you check them out during the next class.

The Unknown Powder

Get Started

Earlier, you had a chance to investigate different kinds of changes.
Changes in matter can be either a **physical change** or a **chemical change**.

A physical change is a change that can be reversed, or changed back.
Changes in state, like an ice cube melting to water, are physical
changes. By putting the melted ice back in the freezer, you can reverse
the change and get the ice cube back.

A chemical change is a change that can't be reversed or is very hard to
reverse. For example, a chemical change occurs when a substance
changes colour, or when a substance gives off bubbles of gas.

With a partner decide which photos on this page show a physical
change, and which show a chemical change.

Think about the changes you observed in Lesson 5. Decide on two examples of physical change and two examples of chemical change you observed.

As a class, create a wall chart that lists the different types of chemical and physical changes you have seen. Leave space on the chart to add more examples you will observe in future.

Now you are beginning to develop an understanding about physical and chemical changes. It's time to investigate four materials that look similar—they are all white powders—but are different. With each powder you will do three different tests. You will use your observations, and your understanding of physical and chemical change, to describe a method for identifying each of the powders.

Materials for each pair:

magnifying glass	baking soda
salt	sugar
corn starch	sheet of paper
vinegar	water
iodine	eye droppers
four test tubes	Popsicle sticks
tablespoon	

Procedure

1 Take a spoonful of each of the following: baking soda, salt, sugar, and starch. Put the powders in separate piles on the sheet of paper and label each pile.

2 Create a table in your notebook. It should look something like the one below.

	Baking soda	Salt	Sugar	Corn starch
Water				
Vinegar				
Iodine				

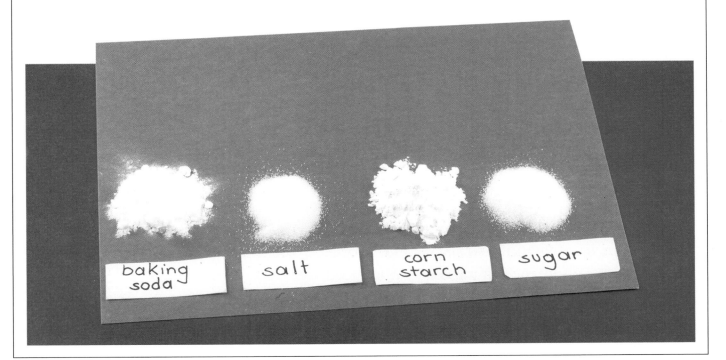

3 Look at each powder under the magnifying glass. Draw what each powder looks like and describe how it looks different from the other powders.

4 Now you will perform the tests. It doesn't matter what order you do them in, as long as you complete all three tests.

Test 1: Water

A Take the four test tubes and half-fill them with water.

B Use a Popsicle stick to add a small amount of each powder to a different test tube.

C Gently shake each test tube.

D Record your observations in the table.

E Clean out the test tubes.

Test 2: Vinegar

A Repeat the steps in Test 1, using vinegar instead of water.

B Record your observations in the table.

C Clean out the test tubes.

Test 3: Iodine

Safety Caution

Wash your hands immediately if you get any iodine on them.

A Use the Popsicle stick to put a small amount of each powder in its own test tube (make sure the test tubes are dry).

B Use the eye dropper to add several drops of iodine to each test tube.

C Record your observations in the table.

D Clean out the test tubes.

5 When you have finished the three tests, look at your table. What test or combination of tests would you use to identify each of the powders? For each test, indicate whether you are observing a physical or chemical change.

The Mysterious Powder

A mysterious white powder was found in the school lunch room. No one is quite sure what it is, but these facts are known.

- The lunch special today was chicken soup, which is thickened by adding corn starch.
- Some baking was done in the cafeteria for parents' night.
- Sweet coffee is the favourite drink of the teachers in the school.

Use this information and your table from the activity you just completed. Determine a method for identifying the mysterious powder.

1. Below are examples of changes in matter. Classify each example as a physical or a chemical change.

   a) a match burning

   b) sugar dissolving in hot coffee

   c) a nail getting rusty

   d) ice forming on a glass of water that is left outside on a cold winter night

2. Describe a chemical change you have observed. Can you think of a way this change could be useful?

3. Describe the change that happens to an egg when it is cooked. Is this a physical or a chemical change?

4. How many changes in matter can you see at home? (Hint: many changes happen in the kitchen just before a meal!) Make a list of all the changes you see. Identify each change as physical or chemical.

Build On What You Know

In your Pocket Book of Change, create a flow chart, like the one started below, to help someone identify an unknown white powder. Use the tests you have used in this activity.

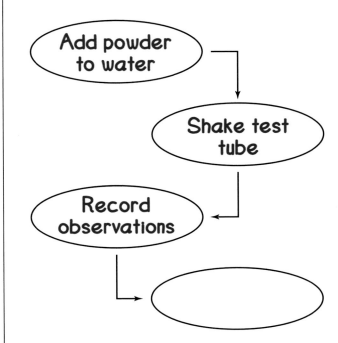

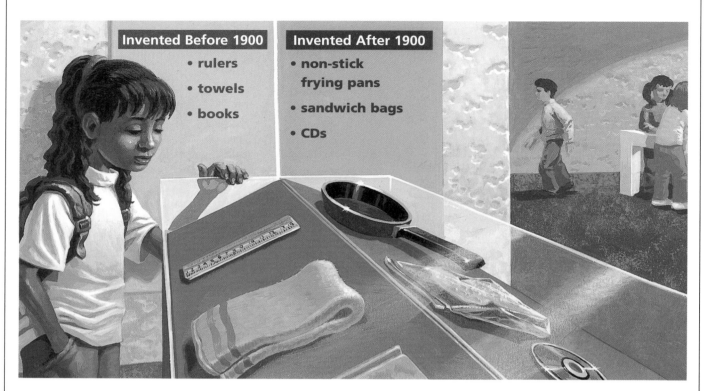

Invented Before 1900	Invented After 1900
• rulers	• non-stick frying pans
• towels	• sandwich bags
• books	• CDs

Get Started

As you enter the Changes Are Useful Room of the museum, you see all sorts of things that look very high tech. You begin to think about what life would be like without technology. You begin to imagine life about a hundred years ago. People were just starting to use cars. Planes were just about to be invented. If you got sick, there were no antibiotics. Listening to music on some kind of device wasn't possible for most people. Television wasn't even invented!

What objects can you find in your home or classroom that were not present 100 years ago? Make a T-chart with one column for materials that were present 100 years ago, and one column for materials that appeared in the last hundred years.

Plastic is one of the most important materials that we use today, but it did not exist 100 years ago. Now we use plastic for everything from artificial limbs, to toys, to containers. We all use plastic all the time! But what is plastic made of and how is it made into so many different shapes?

Plastics are **synthetic**, or human-made, materials made from crude oil. Most of the oil produced in Canada is turned into fuel to run our cars and trucks, and to heat our buildings. But about 2% of the oil we get from the ground is made into **plastic resin**. This resin is then used to produce plastic products.

One reason plastic is so useful is that it can be moulded into almost any shape. Also, plastic is light, and lasts a long time. It is not surprising that plastics can be found almost everywhere we go!

A common way of producing a plastic device in the shape you need is through a process called **extrusion**. During this process, heat and pressure are applied to the plastic resin. This causes the material to melt. The liquid resin is then forced through a specially-shaped nozzle.

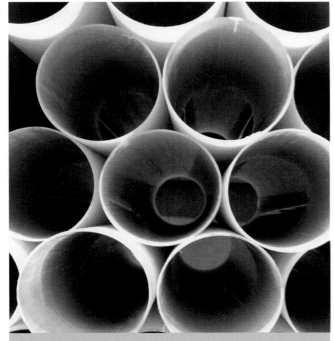

Plastic pipes for plumbing are made by extrusion.

The nozzle has been designed to match the shape of the final product. Once the hot plastic product has been made, or extruded, cold water or air is put over the material to cool it down and make it solid.

In the next activity you will create a model that shows how a plastic shape can be created, shipped, and used.

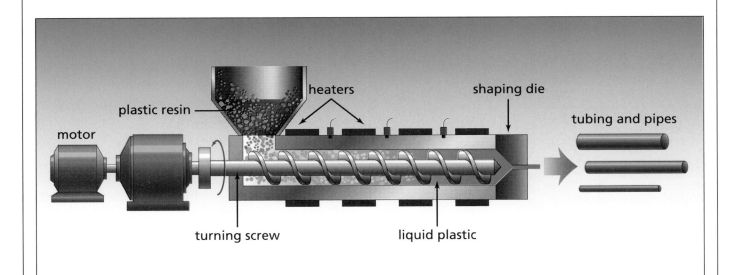

This diagram shows how extrusion is used to make plastic tubing.

Creating a Product with Extrusion

Your group is part of a team that produces plastic materials. You have just received the work order below.

MEMO

To: Extrusion Technology Team

Task: We have just received an order for 10 blue cables 10 cm long, and 2 green pipes at least 5 cm long, both to be made of Sleem.

Special Requirements: Cables and pipes must have smooth surfaces and clean ends. There should be no colour mixing. Both the cables and pipes should be uniform in thickness and size.

Shipping Instructions: The product should be packaged so that it can be dropped from 3 m in the air without being damaged.

Additional Comments: Please determine how much material you start off with. Report how much material was used to make the products, how much was left over, and how much was wasted.

You can make Sleem from the following directions.

Sleem Recipe
500 g of flour
220 g of salt
10 mL of vegetable oil
250 mL of water
food colouring

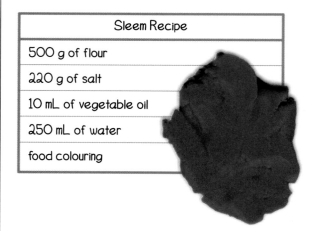

Directions

1. Mix the flour, salt, vegetable oil, and half the water together. Slowly add the rest of the water as needed.

2. Knead the mixture with your hands.

3. Add colour as needed.

4. Knead the mixture to work in the colour until finished.

Materials for each group:
cookie press or pastry bag

two different colours of Sleem

variety of packaging materials like tissue paper, newspaper, small cardboard boxes, etc.

balance

ruler

Procedure

1 Record the mass of the Sleem you are starting with.

2 Using the materials you were given, create the products that are listed in the work order. Let them dry.

3 Once the products are made, record the mass of the products, any materials that were wasted, and the leftover materials. Do the masses add up to give the mass you started with in Step 1?

4 Package your product.

5 Test your packed product to see if it can be shipped without breaking. Stand on a chair and "ship it" to the customer by dropping it from a height of 3 m.

6 Examine your product after the shipping test. Will the customer still accept it?

1. Create a report that explains what you did in this activity. Your report should include any difficulties you encountered. Also, grade your product as being either excellent, good, fair, or needs improvement for the following criteria.

   • My product is the proper colour.

   • My product meets the correct size and number that was required.

   • My product did not dent or break when it was shipped.

   • My product's edges and surfaces are clean and smooth.

   • The mass of Sleem that I started with was equal to the sum of the masses of my product, the waste materials, and the leftover materials.

2. What was the most difficult part of making your product?

3. What would you do differently if you could repeat this activity?

4. What is plastic made from?

Build On What You Know

In your Pocket Book of Change, describe and draw a picture of another product that is made out of plastic using extrusion.

8 The Case of the Extra Mass

Kylee has an ice cube and a cup. Using a balance, she finds the mass of the ice cube to be 12 g. She puts the ice cube in the cup and leaves it on her desk for about 30 min. The whole ice cube has turned into a liquid. She knows this change of state is called melting, and results from heat being transferred from the warm air to the cold ice cube. She places the cup and water back on the scale, and sees that the mass is now 41.5 g. But the ice cube only weighed 12 g at the start! Where did the extra mass come from? Can you help her?

The following activities will help you check your explanation of the situation above. Based on the results you get, you may want to change the explanation you gave.

Changing your explanation doesn't mean that your initial ideas were wrong. You might change your first idea based on new evidence that is discovered. This is an important aspect of science and technology design. Scientists and engineers always adjust their ideas if new evidence calls for a different explanation or design.

Mass During Dissolving

Materials for each group:

sugar

balance

beaker

water

stir stick

paper or paper cup to hold sugar (optional)

Procedure

1 Use the balance to find the mass of the empty beaker. Record the mass.

2 Half-fill the beaker with water. Record the mass of the beaker and the water. What is the mass of the water alone?

3 Use the balance to measure about 20 g of sugar. Record the exact mass of the sugar.

If you use a piece of paper or a paper cup to hold the sugar, remember to subtract the mass of the empty paper or paper cup from the total mass to get the mass of the sugar alone.

4 Add the sugar to the water and stir it until it dissolves.

5 Use the balance to measure the mass of the beaker, water, and sugar. Record the total mass. Calculate the mass of the water and sugar together.

Mass During a Change of State

Materials for each group:

plastic cup

ice cube

balance

elastic band

plastic sandwich bag or other
clear plastic

Procedure

1 Use the materials to design an experiment to find out if the mass of an ice cube changes when it melts. Be sure not to make the same mistake as Kylee did! You may need to measure the mass of the plastic, the elastic band, and the paper cup without the ice cube first.

2 Use the plastic and the elastic band to cover the paper cup while the ice cube is melting. Why is this important? (Hint: Check out Lesson 1 to remind yourself.)

3 Your experiment should answer the following questions.

- What is the mass of the ice cube before it melts?

- What is the mass of the ice cube after it has melted?

Mass During a Chemical Reaction

Materials for each group:

ziplock bag

twist tie

baking soda

teaspoon

vinegar

balance

Safety Caution

Only use the amount of materials as instructed.

Procedure

1 Place 1 teaspoon of baking soda in the corner of the ziplock bag.

2 Seal this corner with a twist tie as shown below.

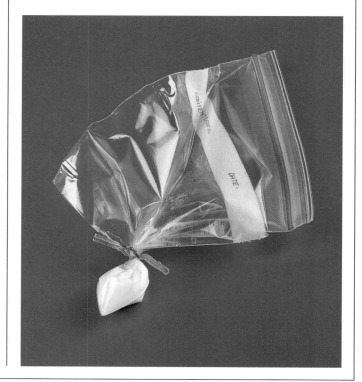

3 Add about 10 mL of vinegar to the bag.

4 Seal the bag so there is very little air remaining in the bag.

5 Determine the mass of the bag and its contents using a balance. Record the mass.

6 Remove the twist tie and mix the two chemicals, leaving the bag closed.

7 Record your observations.

8 After the reaction has stopped, find the mass of the bag and the materials inside by using the balance. Don't forget to include the twist tie on the balance.

9 Record the mass.

10 Look at the mass of the materials before the reaction and compare it to the mass of the materials after the reaction. What conclusion can you make?

Communicate

Write Discuss

1. What is the solution to the Case of the Extra Mass at the beginning of this lesson?

2. Answer these questions for the first activity.

   a) What was the mass of the sugar alone?

   b) What was the mass of the water alone?

   c) What was the total mass when you added the masses of the sugar and water?

   d) What was the total mass of the sugar and water after dissolving?

3. In the second activity, what was the mass of the ice cube before the change of state? What was the mass after it melted?

4. In the third activity, what was the mass of the bag and materials before the reaction? What was the mass after the reaction?

5. What surprised you most about what happened inside the bag in the third activity?

6. For each activity, explain whether you observed a chemical change or a physical change. How do you know?

7. Develop a general rule that explains what happens to mass during a physical or chemical change. Discuss this rule with your class.

Build On What You Know

In your Pocket Book of Change, answer the following question: What happens to the mass of a substance as it changes from a solid to a liquid to a gas? Use a picture to illustrate your thoughts.

9

A Canadian Favourite

Get Started

Try to solve the following riddle about an unknown material. Some of
the first people to try this material were the ancient Greeks, thousands of
years ago. They obtained it as a liquid resin from the bark of a tiny shrub
called the mastic tree. More recently, the explorers of Canada learned
about it from the aboriginal peoples. They tried this material by collecting
the liquid resin from the bark of spruce trees.

Some people didn't like the taste of the resin, so they used sweetened
paraffin wax instead. In the late 1800s, the milky juice from the sapodilla
tree found in Mexico became the favourite choice. What was this
material that everyone wanted to try?

You might not have known it, but the information you read above is a short history of gum! Maybe you thought gum was just something you bought at a store. Believe it or not, a lot of science goes into making a piece of gum. Let's take a tour of a Canadian gum factory and see how they use raw materials to make juicy, delicious gum.

When you see colourful bubblegum balls at your local store, you probably don't think about all the science and technology it takes to make them. It takes five steps and three days to convert the raw materials for gum into the gum balls you see.

1 Melting and Mixing

Originally the basic ingredient in gum was Siamese jelutong, a resin from trees found in Malaysia. This material was used as the base for gum. Now a synthetic, or human made, equivalent is used to make the gum base. This gives the gum its "chewiness." The sweetness of gum comes from all the sugar that is added to it. More than half of every gum ball is just sugar!

The gum base arrives at the factory in large blocks.

Workers break up the gum base and put the small pieces into a kettle. The kettle heats the solid gum base so that it changes state and becomes a liquid. At this point the liquid is thin, like maple syrup. The syrup is filtered to get rid of any impurities. Glucose, a type of sugar, is added at this point as a sweetener. It also makes the gum easier to chew.

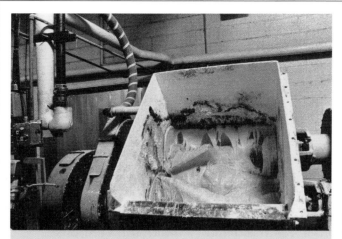

Huge blenders mix sugars and softeners into the liquid base.

After extrusion the long tube of gum travels by conveyor belt to the cutting machine.

The syrup is poured into a large container that looks like a blender. Large blades mix the materials for about half an hour. At this point the gum looks like bread dough, except that it is the colour of ash. During this time, powdered sugars and materials called softeners are added. Softeners are usually glycerin and vegetable oils. They keep the gum moist and easy to chew. At the end of this process, natural or artificial flavours are added.

2 Extrusion

After blending, the gum is extruded into a long snake-like tube. This process is very similar to squeezing toothpaste out of a tube. The thicker the tube, the bigger the gum ball it makes.

A conveyor belt takes the tube of gum to the next machine. During the trip, the gum is cooled to make it firmer than before. This machine looks a little like a set of helicopter blades. The tube of gum is cut into sections and each section is spun around by one of the blades. The spinning action turns the gum into its final round shape. Now you have a gum ball, except that it has an ugly gray colour.

3 It Really is Gross

The round gum balls are now moved by conveyor belt again. This time they are shaken back and forth to make sure they are firm and round. This is called the grossing process. At the end of the process, the gum balls are stored overnight in plastic containers in a cool spot.

These gum balls are going through the grossing process.

The gum balls have to sit overnight before they go through the final stages.

It takes at least two days to turn the raw gum base into the shiny gum balls you see in stores.

4 Almost Finished

After a quiet night, the gum balls are placed in large, spinning kettles. These kettles are full of sugar, food colouring, and liquid flavouring.

Each kettle puts a different colour on the gum ball. The spinning action makes sure each gum ball is painted evenly. The whole process takes about 80 min. The gum ball is almost finished.

5 That Glazed Look

Finally, the gum balls are put in the last kettle of clear liquid colour and beeswax. This glazing process gives the gum a shiny, glossy look. The gum ball is now ready to be packaged and sent to a store near you.

Communicate

Write Present

Make a poster that shows the raw materials used to make gum, and the steps needed to modify these materials to make a gum ball. Display your poster on a wall of your classroom.

Build On What You Know

Choose another common food item and research how it is made. Record what you find out in your Pocket Book of Change.

Finally, the gum balls are given their glossy shine.

10

A Sticky Test

Get Started

The glue being advertised in this ad sounds very strong. The company that makes it is quite sure it works better than any other glue. How can a company be sure of this?

Before the company can make such a claim, they must run tests on their glue. How would they design these tests so they are fair and accurate?

Designing a Fair Test

You probably already know what would be an unfair way to test the glue in the ad. Look at the situations below. Each one describes an unfair test. What is unfair about each situation?

A Glue A is put in the refrigerator for two hours. Glue B is kept at room temperature. Glue A is taken out of the refrigerator and tested against Glue B to see which one does a better job of holding paper on a wall.

B A drop of Glue A is placed between two large blocks. Thirty drops of Glue B are placed between two other similar blocks. Both glues are then tested to see which one holds the blocks together best.

C Ten drops of Glue A are placed on a piece of paper and another piece of paper is placed on top. The glued sheets of paper are allowed to sit overnight and dry. The next day, ten drops of Glue B are placed in a similar way on a piece of paper. A sheet of paper is placed on top, and right away the two glues are tested. The test is to separate the two sheets of paper that have been glued together.

Making Glue

In this activity, you are going to make glue. Then you will design a test that compares your glue to two other types bought from a store.

Materials for each group:

100 mL of skim milk	water
25 mL of vinegar	paper towels
3 g of baking soda	teaspoon
small glass jar	
graduated cylinder	
2 plastic containers (500 mL each)	

Procedure

1 Pour the skim milk into one of the plastic containers. Add the vinegar.

2 Use the spoon to stir the mixture of milk and vinegar. After stirring for about 3 min, you should notice small white lumps forming.

⚠️ Solid white lumps form when the milk and the vinegar react in a chemical change. The solid is called "curd," and the liquid is called "whey."

3 Using the paper towel, make a filter. Hold it over the second plastic container.

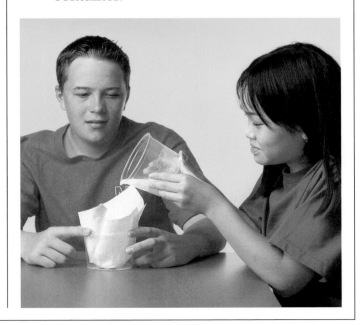

4 Pour $\frac{1}{4}$ of the mixture into the filter. Wait until the liquid passes through the paper towel into the container.

5 Remove the paper towel with the curds in it, and replace it with a new towel. Put the towel with the curds to one side.

6 Pour another equal amount of mixture into a new paper towel filter. Repeat Steps 4 and 5 until all the mixture has passed through a filter.

7 Scrape any remaining curd from the now empty container onto one of the paper towels.

8 Use the rest of your paper towels to dry the curds.

9 Once the curds are as dry as possible, place all of them into the small glass container.

10 Add about 10 mL of water and stir.

11 Add 3 g of baking soda to your new mixture.

Watch your mixture as you add the baking soda. Do you notice anything happening? A chemical reaction is occurring. Can you describe it?

12 Slowly add more water while stirring, until your mixture looks and feels like glue.

Design a Fair Test

Now that you have your glue, you are going to develop a method of testing your glue by comparing it with two brands of store-bought glue.

> ### Safety Caution
>
> Do not use cement glues for your test. Make sure your teacher approves the type of commercial glues you choose.

First you will need to plan a test to find out which glue is the strongest. In your group, discuss and write out your procedure for a fair test. Make sure your procedure includes the following parts.

Procedure

1 Identify your task or the problem you will solve.

2 Decide what kind of fair test will give you an answer to your problem.

3 Identify the factor, or **variable**, that will change during your test. In your procedure, you may be repeating the same steps while changing the type of glue—so the variable might be the type of glue.

All the things that could affect what happens in an experiment are called variables. In any experiment, a scientist wants to test only one thing. But there might be many things that could change the result of an experiment. If you want to test only one of them, you have to keep the other ones the same.

4 Identify other variables that should not change during your test. For example, if you use paper, the same type and amount of paper should be used to test each glue. Can you think of any more variables that should not change?

5 Write out a list of steps that you will follow for your test.

6 List the materials you will need.

7 Make a prediction about what you think will be the result of your tests.

Before you start, have your teacher review your plan. Then perform your test and record your results. Be prepared to present your results to the class.

Communicate

Present Write

1. Which glue was the strongest? How do you know?

2. Which glue was the weakest? How do you know?

3. If you were to repeat this test, what would you do differently next time?

4. What is one thing you learned in this activity that you didn't know before you started?

5. Present the results of your test to the class.

Design Project

Who Stole the Statue?

Get Started

Quick, call the police! Something has been taken from the "It Matters" room in the Interactive Technology Museum of Arts and Sciences. Information is sketchy at this time, but the following is known. The Sleem Statue is missing. It has been replaced by a fake. Only three people have keys to the room: Mr. Sweep, the custodian; Ms. Change, the science and technology demonstrator; and Mr. Beaker, the research consultant.

Suspect one

Mr. Sweep

- He has worked at the museum for many years and probably knows the place better than anyone.

- A white powder was found on his desk. Mr. Sweep says it is just sugar from the donut he had at lunch.

- A star-shaped ring was found in his desk. Police aren't sure what it was used for.

Suspect two

Ms. Change

- She is one of the top presenters in the museum. Her shows are always sold out.

- She was recently heard to complain about how the new Sleem exhibit was taking people away from her shows.

- A white powder was smeared on her lab coat. She says it must have come from one of her demonstrations.

- Ms. Change also said that one of her shows called "Making New Things with Extrusion Technology" uses the same equipment that was used to make the Sleem Statue.

Suspect three

Mr. Beaker

- Mr. Beaker had an excellent science career and was considered to become the next curator of the museum—but he hasn't really done a good job at the museum. He's still looking for the one thing that will make him famous.

- An investigation of his office turned up a funny shaped metal object. The object was circular in shape and had a curved slit.

- White powder was found on the chair in his office. Mr. Beaker refuses to give an explanation for it.

Police were called to the crime scene as soon as it was noticed that the Sleem Statue was a fake. The forgery was spotted when visitors to the museum commented that the curvy beams in the centre of the statue looked messy and uneven. There had also been reports that the statue looked dirtier than usual. Police decided to interview the three people with keys to the It Matters Room. Look at the police reports on these pages for their comments on the three suspects.

Police also have the following information.

- During an interview with the artist who made the original statue, it was discovered that baking soda was spread over the artwork. It seems that this chemical helps absorb dirt and oil left by people touching the statue. According to the artist, this powder would rub off onto the thief. Also, if the statue had been placed on a desk or chair, some powder might be left behind.

- An extrusion engineer indicated that it would be possible to make a copy of the statue using extrusion technology.

Design Project

Samples of the white powder found on all three suspects are available for analysis. Your teacher has obtained these samples for you from the police department.

Your task is to determine which of the three suspects is guilty of stealing the Sleem Statue. You can use any of the information already given, as well as the information you have collected in your Pocket Book of Change. Work with a partner to solve the mystery.

Your answer should identify who stole the statue, explain why you think they took it, and list the evidence you have to support your claim. You should also explain why the other two suspects are not guilty by presenting evidence to prove their innocence.

> **Materials for each pair:**
> Your teacher will provide the materials needed for your analysis of the powders.

Communicate
Write

1. Who stole the Sleem Statue? Provide evidence to support your decision.

2. What evidence did you use to prove that the other two suspects were innocent?

3. Why was it important to understand properties of matter and changes in matter in order to solve this case?